Un-pre-dict-able™

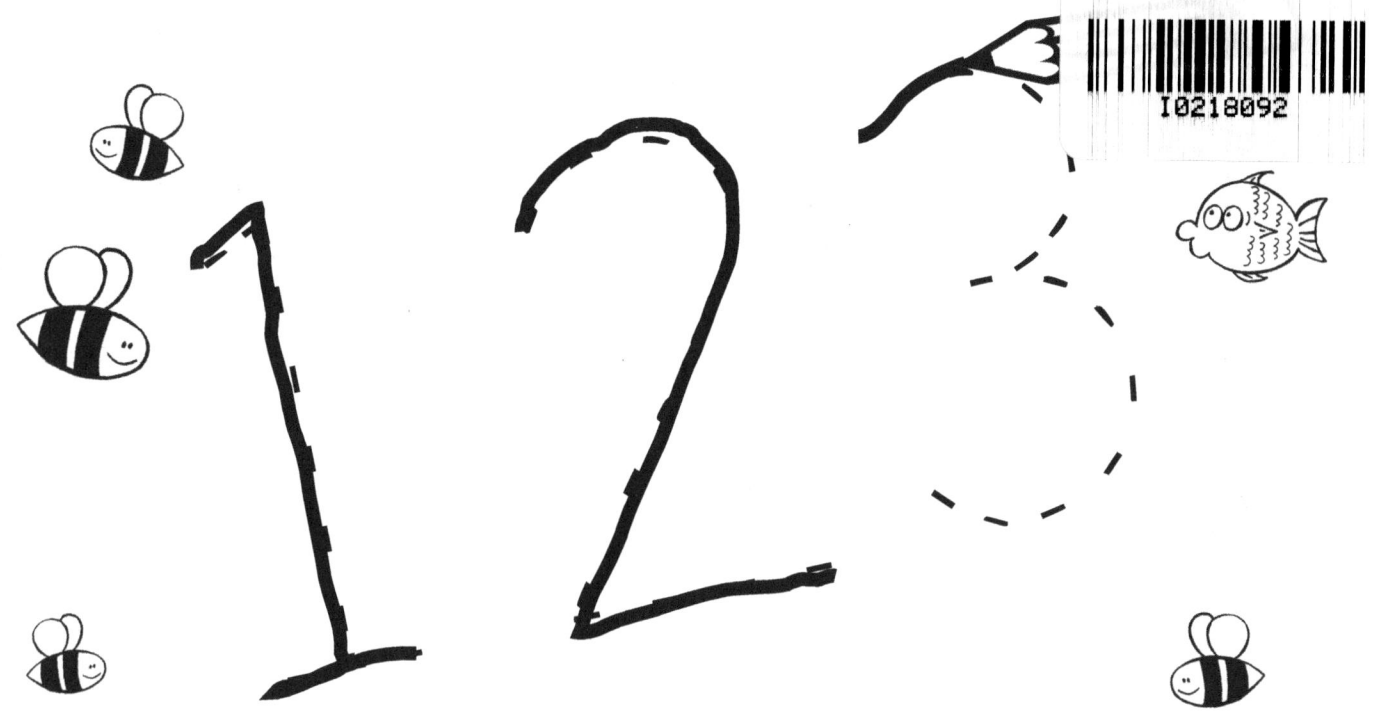

Color, Write and Count

(with some unexpected twists along the way!)

Created by Krysta Bernhardt

Krysta Bernhardt Publishing

www.KrystaBernhardtPublishing.com

Special thanks to David, Gavyn, Owen, my extended family and my friends who have always supported my creative ideas. This little book is dedicated to my inner child who is always saying "you can do this!" even when the grown up me sometimes has other ideas.

ISBN 978-1-7355696-6-6

Copyright © 2021 Krysta Bernhardt Publishing, Zieglerville, PA

www.KrystaBernhardtPublishing.com

All contents, text, design and illustrations created by Krysta Bernhardt

All rights reserved. No part of this book may be reproduced or transmitted in any form or by any means, electronic or mechanical, including photocopying, recording, or by any information storage and retrieval system without the written permission of the author, except where permitted by law.

Trace and write the number below.

Color and count the number of friends below out loud.

Trace and write the word below.

Color the picture, count the ox out loud and write the phrase below.

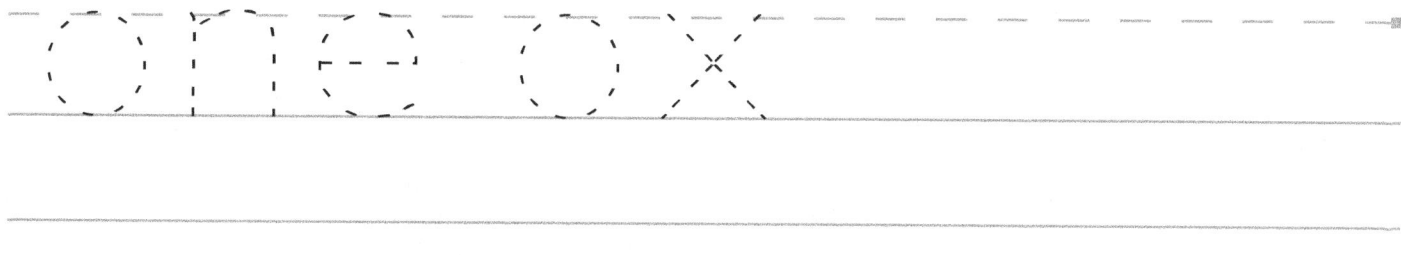

one ox

Trace and write the number below.

Color and count the number of friends below out loud.

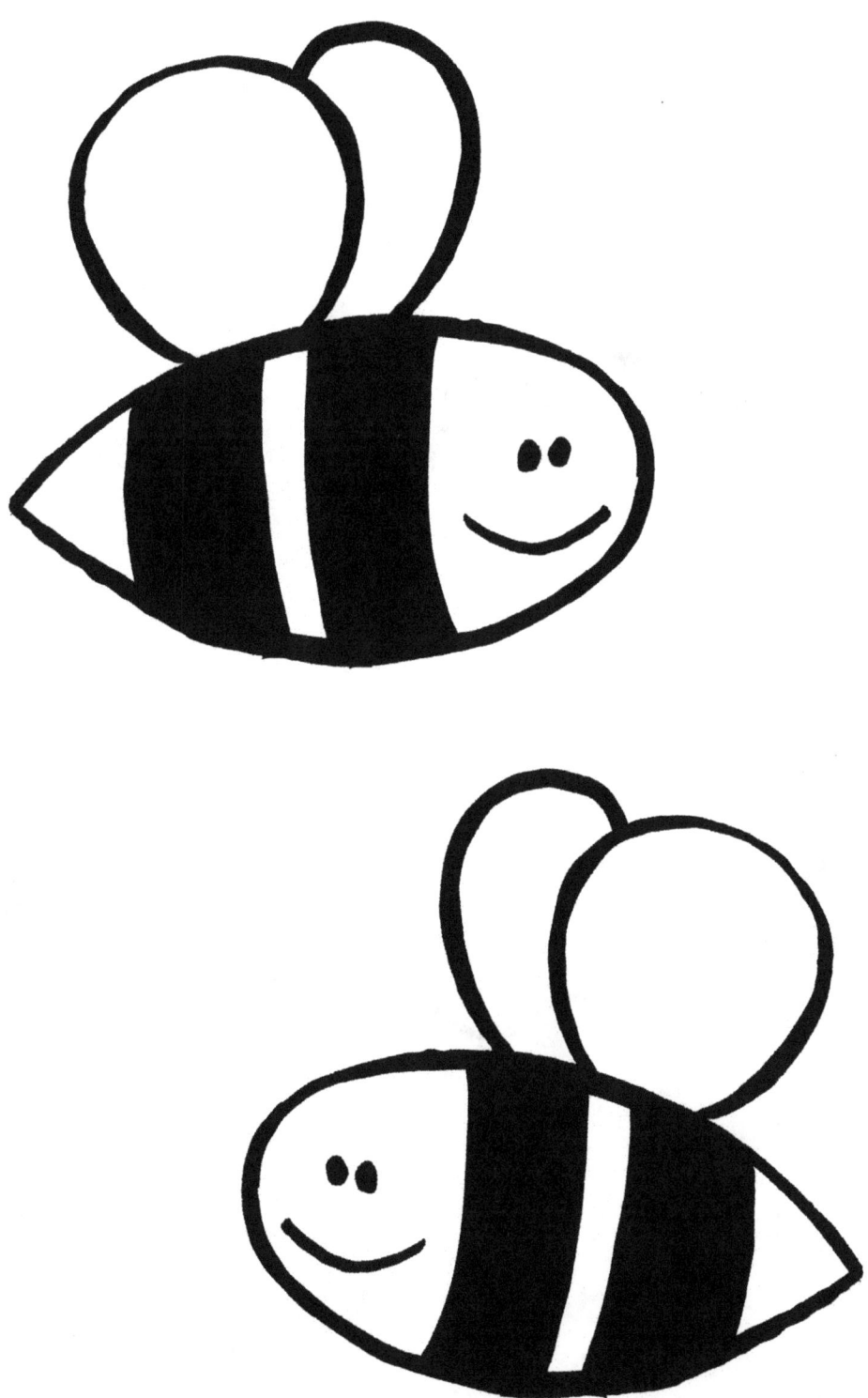

Trace and write the word below.

Color the picture, count the bees out loud and write the phrase below.

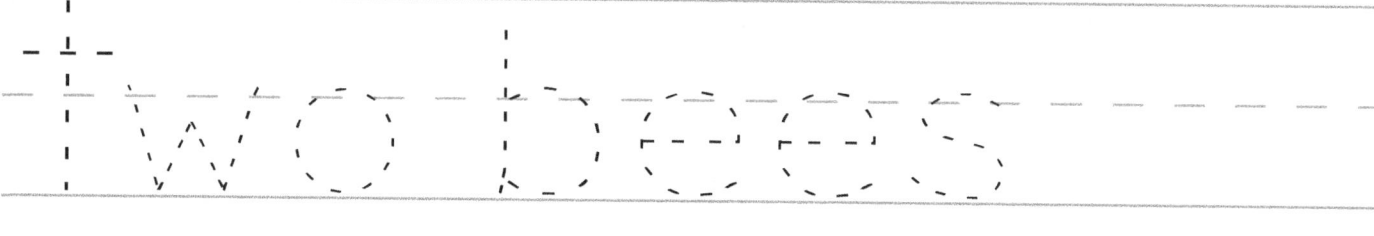

two bees

Trace and write the number below.

Color and count the number of friends below out loud.

Trace and write the word below.

Color the picture, count the egg people out loud and write the phrase below.

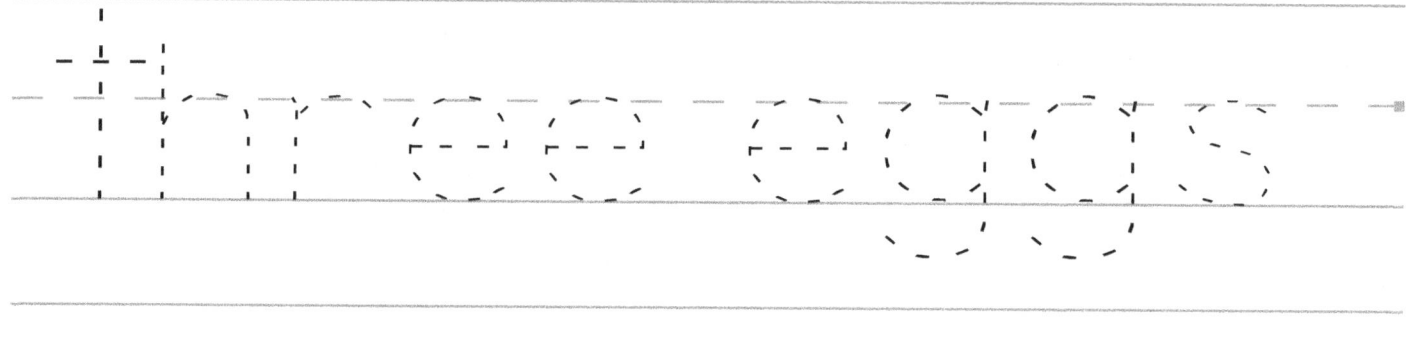

three eggs

Trace and write the number below.

Color and count the number of friends below out loud.

Trace and write the word below.

Color the picture, count the ants out loud and write the phrase below.

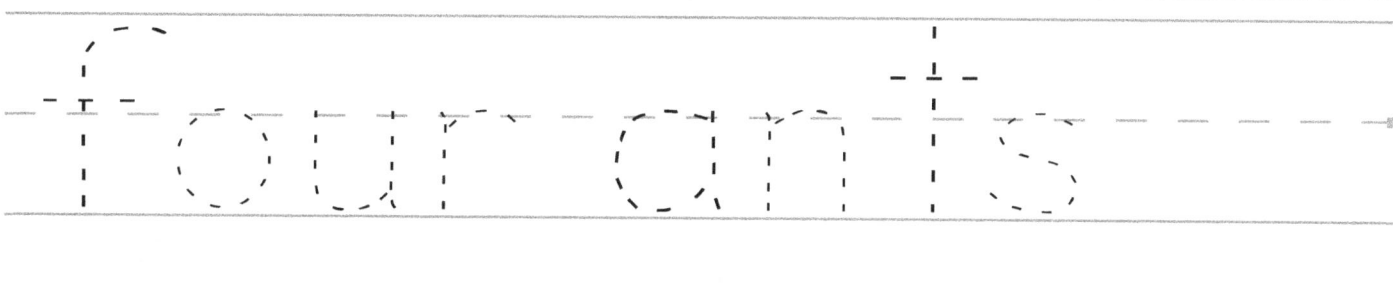

four ants

Trace and write the number below.

Color and count the number of friends below out loud.

Trace and write the word below.

Color the picture, count the cats out loud and write the phrase below.

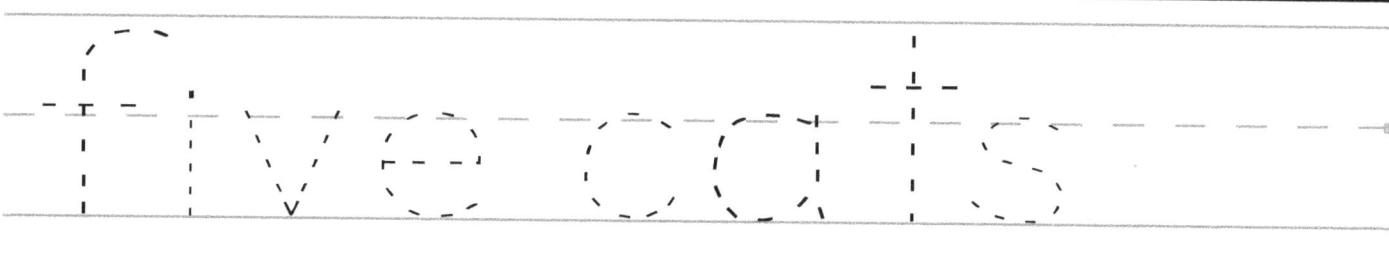

five cats

Trace and write the number below.

Color and count the number of friends below out loud.

Trace and write the word below.

Color the picture, count the dancing dogs out loud and write the phrase below.

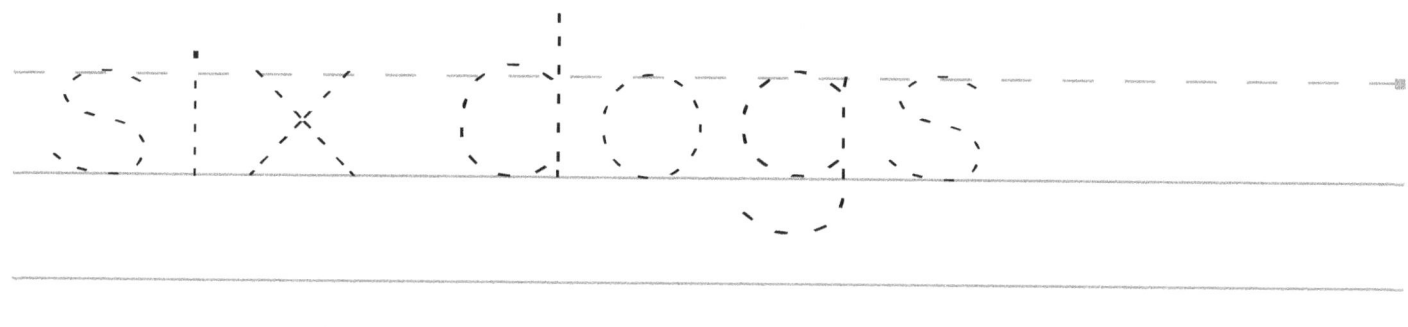

six dogs

Trace and write the number below.

Color and count the number of friends below out loud.

Trace and write the word below.

seven

seven

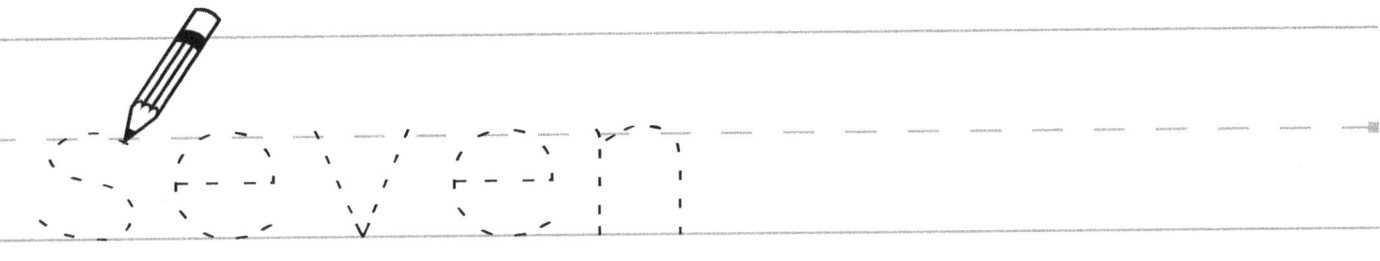

Color the picture, count the pigs out loud and write the phrase below.

seven pigs

Trace and write the number below.

Color and count the number of friends below out loud.

33

Trace and write the word below.

eight

eight

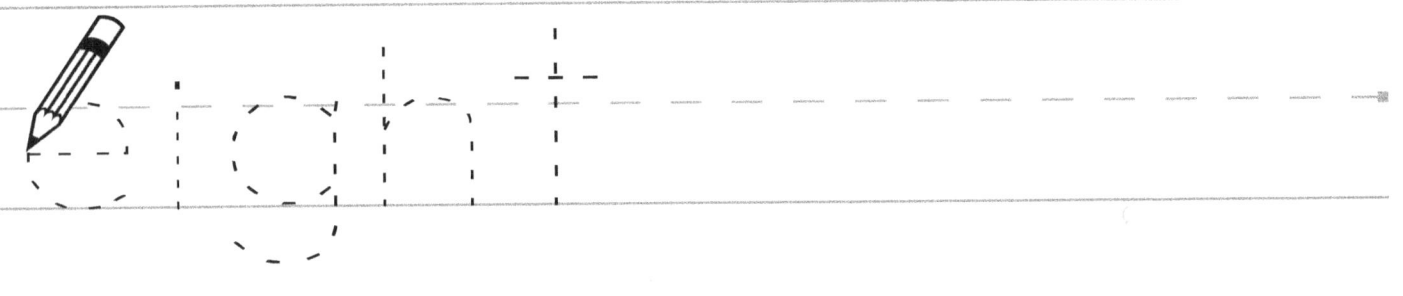

Color the picture, count the fish out loud and write the phrase below.

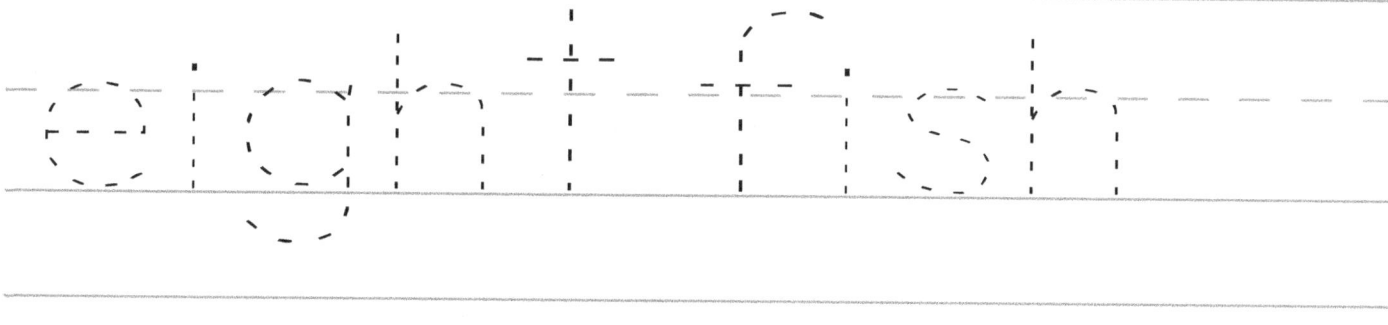

eight fish

Trace and write the number below.

Color and count the number of friends below out loud.

Trace and write the word below.

nine

nine

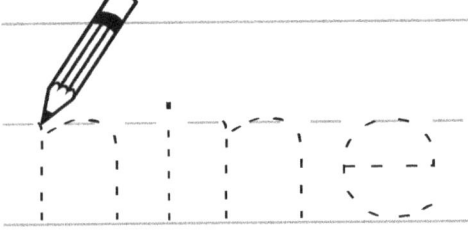

Color the picture, count the owls out loud and write the phrase below.

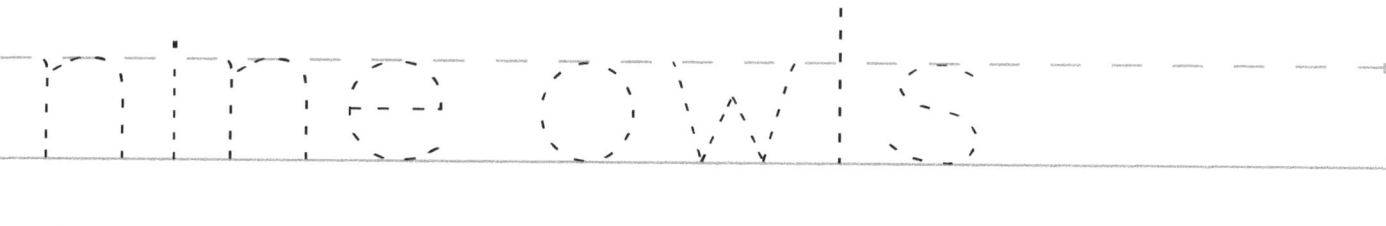

nine owls

Trace and write the number below.

Color and count the number of friends below out loud.

Trace and write the word below.

Color the picture, count the bees out loud and write the phrase below.

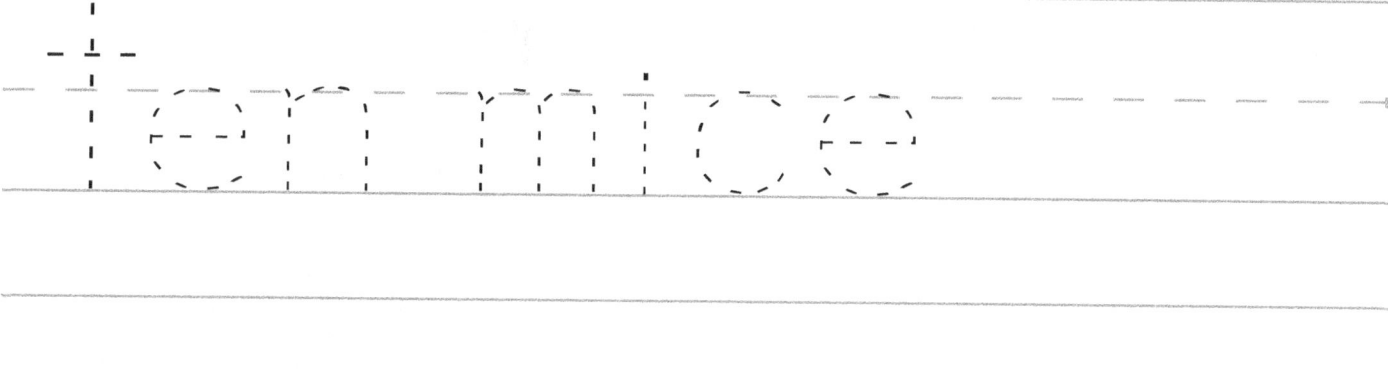

ten mice

 # CONGRATULATIONS!

_____
Your name

has completed

Keep practicing and HAPPY COUNTING!

Krysta Bernhardt lives and creates in a little house that she shares with her kids, her husband and her two fat cats. Krysta loves to teach kids, draw, paint, sing, dance, make music and create. Her favorite books have always had big, bold lines and simple, imperfect characters. She believes that if you have an idea, even an imperfect one, it deserves to be created and shared! Keep an eye out for more of Krysta's Un-pre-dict-a-ble and imperfect creations!

Krysta Bernhardt Publishing

If you enjoyed your Un-pre-dict-a-ble adventure, please consider leaving a review on Amazon so that other young learners can find out about and enjoy these books! Thank you so much for your support! – Krysta

Find more books at www.KrystaBernhardtPublishing.com

www.ingramcontent.com/pod-product-compliance
Lightning Source LLC
Chambersburg PA
CBHW081024040426
42444CB00014B/3338